①

②

③

④

⑤

⑥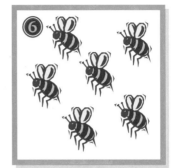

S

O

G

N

E

P

What is full of holes and can still hold water?

① with   ② with   ③ with   ④ with   ⑤ with   ⑥ with

A ___ ___ ___   ___ ___

① 1

MATCH THE SETS.

What kind of key can you eat?

① with ② with ③ with ④ with ⑤ with ⑥ with

A ___ ___ ___ ___ ___ ___

T

Y

U

R

E

K

ADDITION

Match.

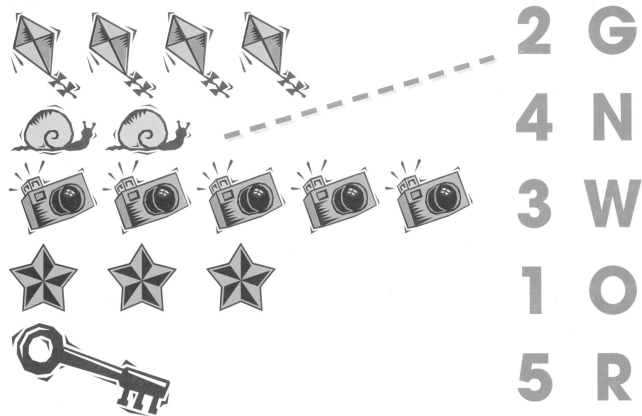

2	G
4	N
3	W
1	O
5	R

What word do we always say wrong?
Write the letter above the number to find out.

___ ___ ___ ___ ___
3   5   1   4   2

Draw enough of each shape to make 5.

Match the sets and solve the puzzle.

Match.

OOOOOOO — 6 A

XXXXXXX — 7 O

OOOOOO — 8 C

XXXXXXXXX — 9 I

OOOOOOOOOO — 10 E

What game is played with Xs and Os?
Write the letter above the number to find out.

T ___ ___ - T ___ ___ - T ___ ___

9 8 6 8 7 10

Draw enough Xs to make 10.

O O O O O O

O O O O O O O O

O O O O O

O O O O O O O O O

O O O O O O O

Trace the numbers.

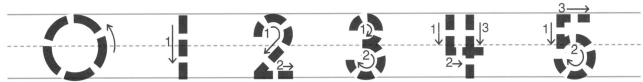

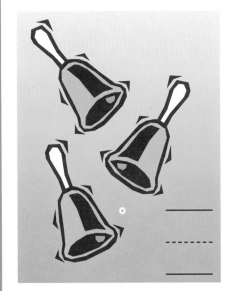

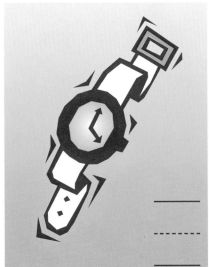

Write numbers from 0 to 5 in order.

Trace the numbers.

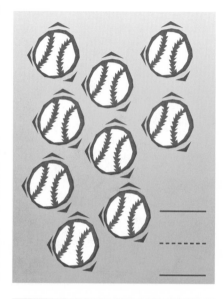

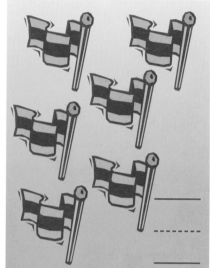

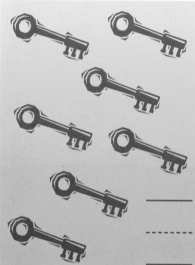

Write numbers from 6 to 10 in order.

Circle

	Set 1	Set 2		Set 1	Set 2
			1	Ⓐ	G
			2	U	A
			3	S	T
			4	O	R
			5	L	M
			6	O	I
			7	N	B
			8	I	E
			9	A	L
			10	E	S

What 10-letter word starts with "gas"?
Write the circled letters above the numbers.

A _ _ _ _ _ _ _ _ _
1 2 3 4 5 6 7 8 9 10

7

ADDITION

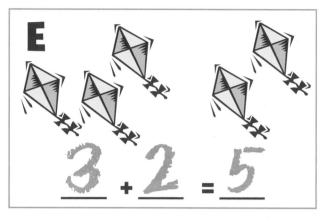

E

$3 + 2 = 5$

___ + ___ = ___

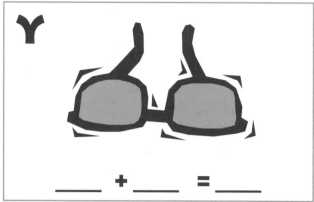

Y

___ + ___ = ___

B

___ + ___ = ___

U

___ + ___ = ___

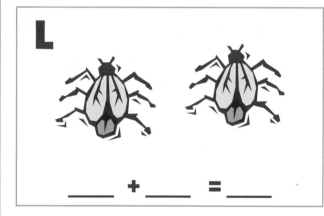

L

___ + ___ = ___

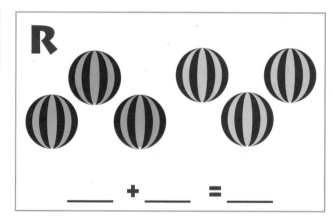

R

___ + ___ = ___

Write the letter in each box above the correct sum. Answer this riddle:
When is a color good to eat?

When it's a

___ ___ ___ ___ ___ ___ ___ ___ ___ !
4   2   3   5   4   5   6   6   1

8

ADDITION

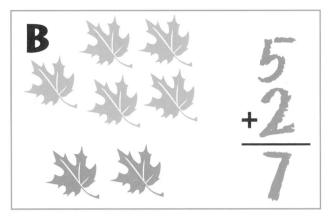

B

$$\begin{array}{r} 5 \\ + 2 \\ \hline 7 \end{array}$$

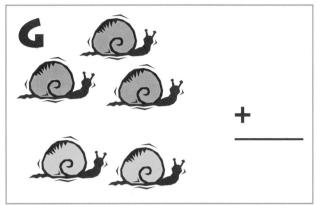

G

$+$ ___

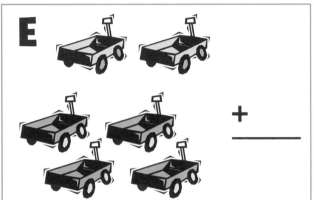

E

$+$ ___

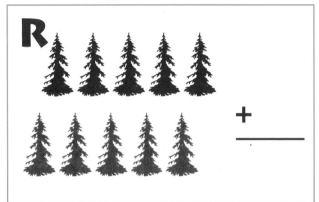

R

$+$ ___

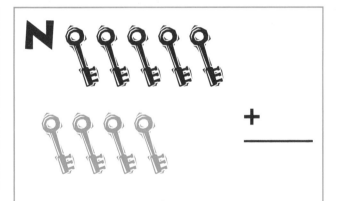

N

$+$ ___

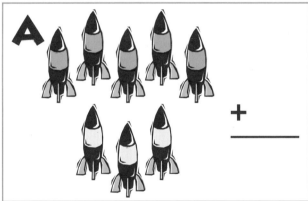

A

$+$ ___

Write the letter in each box above the correct sum. Answer this riddle:
When is a color good to eat?

When it's a

___ ___ ___ ___ ___   ___ ___ ___ ___
 5   10   6   6   9   7   6   8   9

9

SOLVE THE PROBLEMS, THEN COLOR.

Color sums 0 – 7 yellow; 8 green; 9 blue; 10 red

6 + 1	2 + 4	4 + 0	3 + 3	1 + 3	0 + 4	5 + 2
2 + 2	2 + 5	3 + 2	7 + 0	1 + 5	3 + 4	4 + 2
4 + 3	2 + 8	9 + 0	5 + 3	1 + 9	2 + 7	1 + 6
0 + 7	6 + 3	3 + 7	6 + 2	5 + 4	5 + 5	1 + 4
3 + 3	7 + 1	0 + 8	4 + 4	8 + 0	1 + 7	3 + 2
2 + 5	8 + 2	8 + 1	2 + 6	6 + 4	0 + 9	1 + 0
1 + 6	4 + 5	9 + 1	3 + 5	3 + 6	7 + 3	3 + 1

10

Ben had 5 toy cars.

Eric had 4 toy cars.

How many toy cars in all?

$5 + 4 = 9$

Janet bought 7 sea shells.

David bought 2.

How many sea shells did both children have in all?

Josh bought 6 apples.

Jenny bought 4 apples.

How many apples were there in all?

Marlon bought 4 footballs.

Larry bought 3 footballs.

How many footballs did they have in all?

Chris had 3 pencils.

Jill had 6 pencils.

How many pencils did Chris and Jill have all together?

James took home 7 books for studying.

Sarah took 3 books home for studying.

How many books did they both take home?

Rita saw 4 ducks.

Ron also saw 4 ducks.

How many ducks did both the children see?

Carol has 3 fish in her fish bowl.

Lori has 5 fish.

How many fish do both the girls have together?

11

Write the missing addend.
Write its matching letter.

3	2	5	9	7	1	6	8	4
A	C	K	L	O	R	T	U	W

What should you do for a broken toe?

4 + 2 6	+ 3 6	1 + __ 10	+ 0 9
C			

7 + __ 10

+ 3 9	2 + __ 9	+ 4 8

2 + __ 8	+ 6 7	2 + __ 10	+ 7 9	4 + __ 9

12

ADDITION

Write the missing numbers.

0 [ ] 2 [ ] [ ] 5 [ ] 7 [ ] [ ] 10

Write an equation.

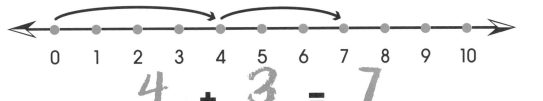

$$\underline{4} + \underline{3} = \underline{7}$$

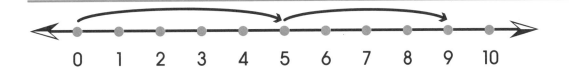

$$\underline{} + \underline{} = \underline{}$$

$$\underline{} + \underline{} = \underline{}$$

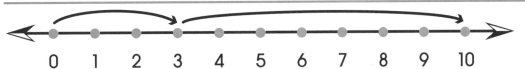

$$\underline{} + \underline{} = \underline{}$$

$$\underline{} + \underline{} = \underline{}$$

$$\underline{} + \underline{} = \underline{}$$

13

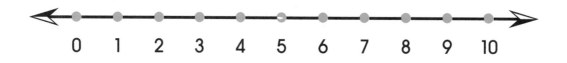

0 1 2 3 4 5 6 7 8 9 10

>
greater than

<
less than

Write number sentences:

Sample:

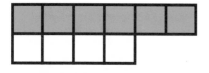

6 is greater than 4
4 is less than 6

6 > 4
4 < 6

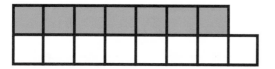

7 ◯ 8
8 ◯ 7

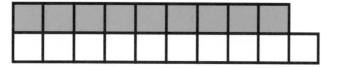

9 ◯ 10
10 ◯ 9

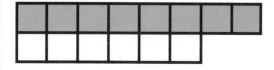

8 ◯ 6
6 ◯ 8

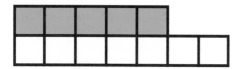

5 ◯ 7
7 ◯ 5

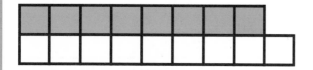

8 ◯ 9
9 ◯ 8

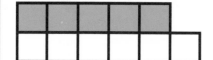

5 ◯ 6
6 ◯ 5

14

ADDITION

9th	8th	10th	4th	3rd	6th	5th	7th	1st	2nd

Put the letters in the correct blank.

seventh letter
O

third letter
T

fifth letter
H

ninth letter
K

eighth letter
R

fourth letter
C

first letter
P

second letter
I

tenth letter
S

sixth letter
F

What weather is best for the farmer to make hay?
When it is raining …

 ___ ___ ___ ___ ___ ___ ___ ___ ___ ___

15

COMPLETE THE PROBLEMS.

+ 3	
7	10
5	
0	
6	
4	
2	
3	
1	

+ 4	
2	
4	
3	
0	
1	
6	
5	

+ 2	
8	
10	
7	
9	
6	
5	
2	
3	

+ 1	
2	
6	
1	
5	
8	
3	
0	
7	

+ 5	
6	
4	
2	
7	
5	
3	
1	
0	

+ 6	
0	
5	
2	
6	
4	
1	
7	
3	

16

ADDITION

6	4	10
5	2	
11		

5	4	
3	2	

7	3	
6	2	

4	6	
3	5	

5		10
	3	6
8		

	5	9
1		5
	9	

ADDITION

Complete.

	Table Form TENS	ONES	Expanded Form	Standard Form
	1	0	10 + 0	10
	1	1	10 + ☐	☐
	1	2	10 + ☐	☐
	1	3	10 + ☐	☐
	☐	4	10 + ☐	☐
	1	☐	10 + ☐	☐
	1	7	☐ + ☐	☐
	☐	☐	10 + 8	☐
	☐	☐	☐ + ☐	19

18

ADDITION

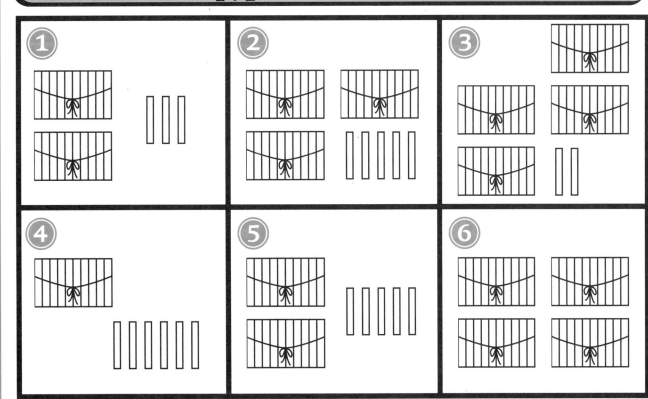

I 35   S 40   G 23

T 25   A 42   N 16

What are the largest kinds of ants?

① with   ② with   ③ with   ④ with   ⑤ with   ⑥ with

G ___ ___ ___ ___ ___

ADDITION

COMPLETE THE PROBLEMS.

6 tens and 4 ones = __60__ + __4__ = __64__

5 tens and 8 ones = _____ + _____ = _____

2 tens and 3 ones = _____ + _____ = _____

1 ten and 9 ones = _____ + _____ = _____

4 tens and 7 ones = _____ + _____ = _____

3 tens and 2 ones = _____ + _____ = _____

9 tens and 6 ones = _____ + _____ = _____

8 tens and 5 ones = _____ + _____ = _____

7 tens and 1 one = _____ + _____ = _____

59 means __5__ tens and __9__ ones.

26 means _____ tens and _____ ones.

12 means _____ ten and _____ ones.

45 means _____ tens and _____ ones.

34 means _____ tens and _____ ones.

93 means _____ tens and _____ ones.

86 means _____ tens and _____ ones.

78 means _____ tens and _____ ones.

40 means _____ tens and _____ ones.

20

FOLLOW DIRECTIONS TO COLOR THE PUZZLE.

Color number names less than 50 in RED.
Color number names more than 50 in BLUE.

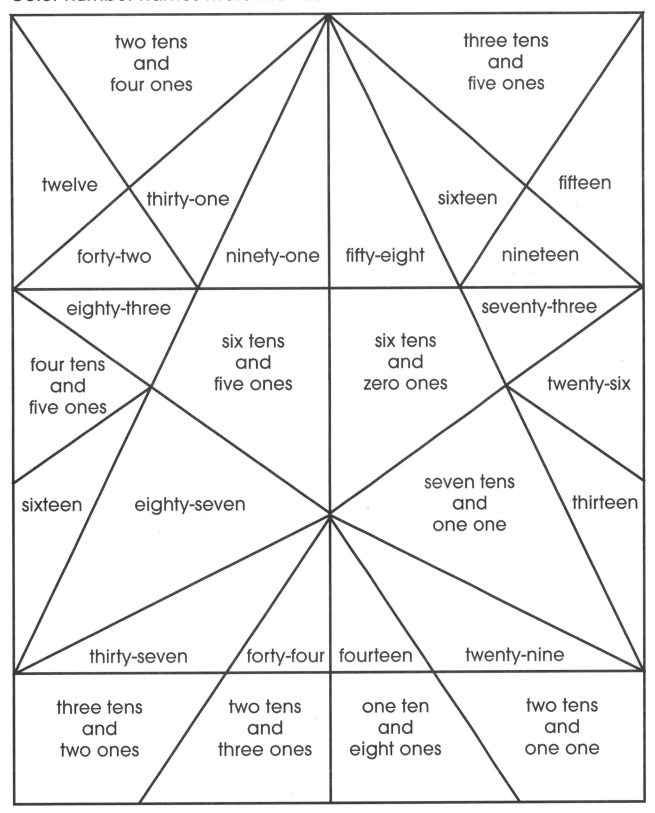

two tens and four ones

three tens and five ones

twelve

thirty-one

sixteen

fifteen

forty-two

ninety-one

fifty-eight

nineteen

eighty-three

seventy-three

four tens and five ones

six tens and five ones

six tens and zero ones

twenty-six

sixteen

eighty-seven

seven tens and one one

thirteen

thirty-seven

forty-four

fourteen

twenty-nine

three tens and two ones

two tens and three ones

one ten and eight ones

two tens and one one

21

ADDITION

Complete to count by 3s to 30.

1   2   ③   4   5   ◯   7   8   ◯   10   11   ◯   13   14   ◯

16   17   ◯   19   20   ◯   22   23   ◯   25   26   ◯   28   29   ◯

Complete to count by 4s to 40.

1   2   3   ④   5   6·   7   ◯   9   10   11   ◯   13   14   15

◯   17   18   19   ◯   21   22   23   ◯   25   26   27   ◯   29   30

31   ◯   33   34   35   ◯   37   38   39   ◯

Count by 3s from 3 to 30.
Connect the ● s.

Count by 4s from 4 to 40.
Connect the + s.

Count by 5s from 5 to 50.
Connect the ★ s.

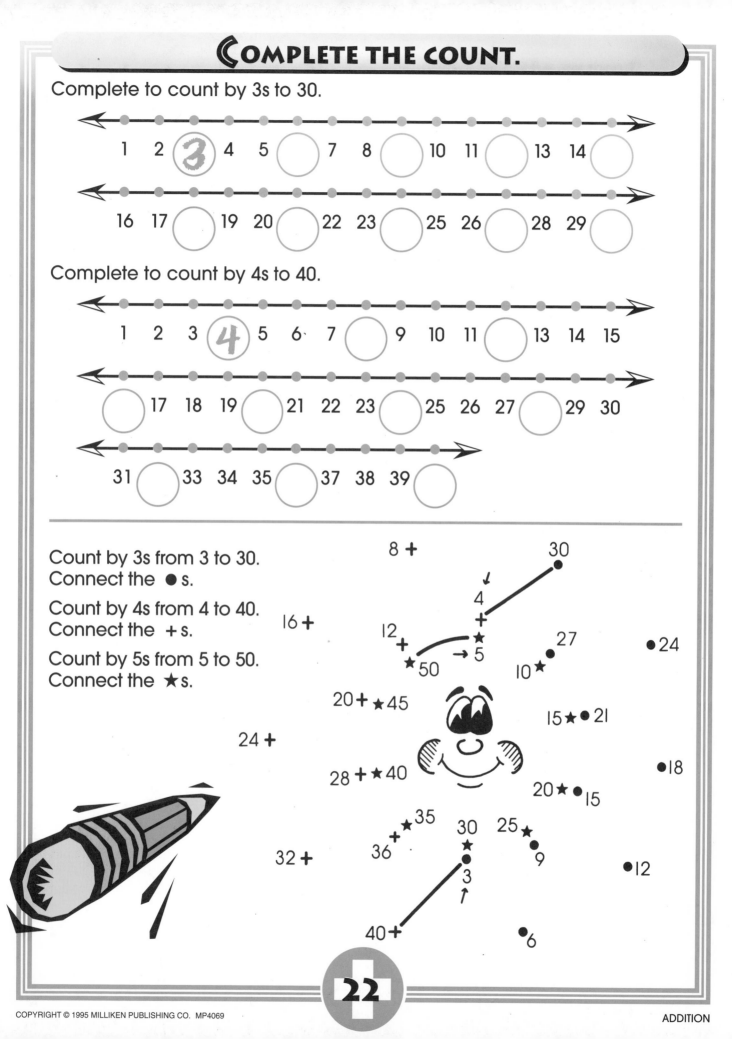

8 +

30

16 +

4
↓
+
12 +   ★
★ 50   → 5

27
10 ★

● 24

20 + ★ 45

15 ★ ● 21

24 +

● 18

28 + ★ 40

20 ★ ● 15

★ 35
36   + ★

30
★
25   ★
● 9

32 +

● 12

3
↑

40 +

● 6

ADDITION

Answer Key

for

Milliken's

Math Workbook

on

ADDITION

If you wish for the child to correct his or her own work, leave this answer key bound in the workbook for easy reference.

However, if you wish for someone else to correct the child's work, remove this answer key and keep it separately from the workbook. The key should separate easily, but if you feel it may tear, bend the staples out, slip the key out and bend the staples back to keep the workbook intact.

Answers are set up in order of page number, by rows and columns, reading left to right.

PAGE 1:
   Sponge

PAGE 2:
   Turkey

PAGE 3:
   Kites: 4
   Cameras: 5
   Stars: 3
   Key: 1
   Wrong
   3 circles
   1 diamond
   4 triangles
   3 stars
   2 hearts

PAGE 4:
   1st row of Xs: 8
   2nd row of Os: 6
   2nd row of Xs: 10
   3rd row of Os: 9
   Tic-Tac-Toe
   4 Xs
   2 Xs
   5 Xs
   1 X
   3 Xs

PAGE 5:
   3, 1, 4
   5, 0, 2
Children should draw the numbers according to the directions on the page.

PAGE 6:
   9, 10, 6
   7, 8, 9
Children should draw the numbers according to the directions on the page.

PAGE 7:
   Automobile

PAGE 8:
   1 + 0 = 1
   1 + 3 = 4; 2 + 1 = 3
   1 + 1 = 2; 3 + 3 = 6
   Blueberry

PAGE 9:

$$\begin{array}{r} 3 \\ + 2 \\ \hline 5 \end{array}$$

$$\begin{array}{r} 2 \\ + 4 \\ \hline 6 \end{array} \qquad \begin{array}{r} 5 \\ + 5 \\ \hline 10 \end{array}$$

$$\begin{array}{r} 5 \\ + 4 \\ \hline 9 \end{array} \qquad \begin{array}{r} 5 \\ + 3 \\ \hline 8 \end{array}$$

   Green Bean

PAGE 10:

7	6	4	6	4	4	7
4	7	5	7	6	7	6
7	10	9	8	10	9	7
7	9	10	8	9	10	5
6	8	8	8	8	8	5
7	10	9	8	10	9	1
7	9	10	8	9	10	4

Children should color the regions according to the directions on the page.

PAGE 11:
   7 + 2 = 9
   6 + 4 = 10; 4 + 3 = 7
   3 + 6 = 9; 7 + 3 = 10
   4 + 4 = 8; 3 + 5 = 8

PAGE 12:
   3, 9, 9, 3
   6, 7, 4, 6, 1, 8, 2, 5
   Call a tow truck.

PAGE 13:
   5 + 4 = 9
   6 + 3 = 9
   3 + 7 = 10
   1 + 7 = 8
   4 + 6 = 10

PAGE 14:
   7 is less than 8;
    8 is greater than 7
   9 is less than 10;
    10 is greater than 9
   8 is greater than 6;
    6 is less than 8

PAGE 14 (continued):
   5 is less than 7;
    7 is greater than 5
   8 is less than 9;
    9 is greater than 8
   5 is less than 6;
    6 is greater than 5
   <
   >
   <
   >
   >
   <
   <
   >
   <
   >
   <
   >

PAGE 15:
   Children should match the train cars with their position in the train according to the directions on the page.
   Pitchforks

PAGE 16:
   + 3: 8, 3, 9, 7, 5, 6, 4
   + 4: 6, 8, 7, 4, 5, 10, 9
   + 2: 10, 12, 9, 11, 8, 7, 4, 5
   + 1: 3, 7, 2, 6, 9, 4, 1, 8
   + 5: 11, 9, 7, 12, 10, 8, 6, 5
   +6: 6, 11, 8, 12, 10, 7, 13, 9

PAGE 17:
   Left column:
    7, 6, 17
   10, 8, 13, 5, 18
    5, 3, 8, 16
   Right column:
    9, 5, 8, 6, 14
    10, 8, 7, 11, 18
    4, 4, 5, 14

PAGE 18:
   1 11
   2 12
   3 13
   1 4 14
   5 5 15

1 ten, 7 ones; 10   7   17
1 ten, 8 ones; 1   8   18
1 tens, 9 ones; 1   9   10   9

PAGE 19:
Giants

PAGE 20:
50 + 8 = 58
20 + 3 = 23
10 + 9 = 19
40 + 7 = 47
30 + 2 = 32
90 + 6 = 96
80 + 5 = 85
70 + 1 = 71
2   6
1   2
4   5
3   4
9   3
8   6
7   8
4   0

PAGE 21:
Children should color the regions according to the directions on the page.

PAGE 22:
6   9   12   15
18   21   24   27   30
8   12
16   20   24   28
32   36   40
Children should connect the dots according to the directions on the page.

PAGE 23:
2, 4, 6, 8, 10, 12, 14, 16,
18, 20, 22, 24, 26, 28,
30, 32, 34, 36, 38, 40,
42, 44, 46, 48, 50
5, 10, 15, 20, 25, 30, 35,
40, 45, 50
Children should connect the dots according to the directions on the page.

PAGE 24:
6   9   60 + 9   69
4   8   40 + 8   48
79   78   97   73   79
76   87   68   97   98

PAGE 25:
47   79   57   26   98
46   79   98   55   98   87
98   47   79
66   87   98   98   87   69
When it gets to the bottom.

PAGE 26:
79
46 + 22 = 68
25 + 12 = 37
37 + 41 = 78
63 + 34 = 97
54 + 35 = 89

PAGE 27:
>   >   <
>   <   <
>   <   <
>   >   >
>   <   <

PAGE 28:
90   60   90   50
50   70   90   90   80

$$\begin{array}{r} 40 \\ + 30 \\ \hline 70 \end{array}$$

$$\begin{array}{r} 20 \\ + 60 \\ \hline 80 \end{array} \qquad \begin{array}{r} 10 \\ + 50 \\ \hline 60 \end{array}$$

PAGE 29:
69   89   28   86   47   97
66   41   37   58   97   55
77   98   49   46   99   87

PAGE 30:
96   68   28
36   66   78
46   95   52   39   52   98
88   80   98   89   88   98
66   97   77
57   59   36

PAGE 31:
13   13   15   9   9   8
33   95   96
84   89   68
72   92   95   94   64   93

PAGE 32:
92   80   62   71   60
81   81   53   64   91

PAGE 33:
1.   61
2.   81
3.   90
4.   75
5.   90
6.   90

PAGE 34:
64   80   92   82   80   70
32   70   52   60   81   62
93   83   54   93   80   94
84   43   62   81   47   81

PAGE 35:
Reading diagonally:
80   81   90
32   70   42   30
72   63   93   80   91
80   92   81   71   38
81   73   50

PAGE 36:
$$\begin{array}{r} 18 \\ + 45 \\ \hline 63 \end{array} \qquad \begin{array}{r} 18 \\ + 18 \\ \hline 36 \end{array}$$

$$\begin{array}{r} 38 \\ + 56 \\ \hline 94 \end{array} \qquad \begin{array}{r} 24 \\ + 18 \\ \hline 42 \end{array} \qquad \begin{array}{r} 45 \\ + 24 \\ \hline 69 \end{array}$$

$$\begin{array}{r} 45 \\ + 45 \\ \hline 90 \end{array} \qquad \begin{array}{r} 24 \\ + 56 \\ \hline 80 \end{array} \qquad \begin{array}{r} 38 \\ + 18 \\ \hline 56 \end{array}$$

$$\begin{array}{r} 56 \\ + 18 \\ \hline 74 \end{array} \qquad \begin{array}{r} 38 \\ + 45 \\ \hline 83 \end{array} \qquad \begin{array}{r} 24 \\ + 24 \\ \hline 48 \end{array}$$

PAGE 37:
1. 32
2. 78
3. 25
4. 61
5. 40
6. 79

PAGE 38:

Each star is read clockwise beginning at the top:

Top row:
41 50 85 32 53 62
94 74
25 41 77 30 63 14
21 54

Center:
64 71 81 90 95 97
51 58

Bottom row:
51 77 42 65 27 40
60 22
50 97 40 55 71 88
65 75

PAGE 39:
47 36
27 81 45
30 42 38
76 99 42
95 69 82

PAGE 40:
Left column:
27, 65, 70, 22, 92
40, 55, 64, 31, 95
53, 76, 66, 63, 129
Right column:
45, 41, 63, 23, 86
39, 60, 64, 35, 99
103, 54, 84, 73, 157

PAGE 41:
800 600 700 900 800
900 500 800 900 300
6   9   5
600 + 200 = 800
400 + 300 = 700
100 + 200 = 300

PAGE 42:
881 977 598 873 958
999 985
959 599 998 769 576
777 557

PAGE 43:
851 859 597 861 675
684
638 364 687 871 943
678
851 849 858 780 473
791
968 865 669 982 881
762

PAGE 44:
1. 693
2. 990
3. 126
4. 147
5. 606
6. 285

Write the numbers from 1 to 50.

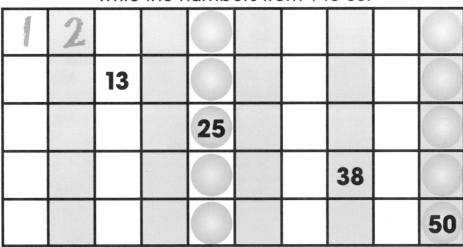

To count by 2s from 2 to 50:
Write only the numbers in the shaded boxes in order.

___ ___ ___ ___ ___ ___ ___ ___ ___ ___

___ ___ ___ ___ ___ ___ ___ ___ ___ ___

___ ___ ___ ___ ___

To count by 5s from 5 to 50:
Write only the numbers in the circles in order.

___ ___ ___ ___ ___ ___ ___ ___ ___ ___

Count by 2s to 50. Connect the ●s. Count by 5s to 50. Connect the ★s.

```
     2●    4●    6●    8●   10●   12●   14●   16●   18●   20●
40★                                                        ★5
        35★                                      22●
                    30★              24●
45★                              26●                    ★10
              28●                            25★
        30●                                      20★
50★                                                    ★15
  32    34    36    38    40    42    44    46    48    50
```

ADDING TENS AND ONES.

Complete. Add.

	Table Form TENS	ONES	Expanded Form	Standard Form
	3	4	30 + 4	34
	+ 2	4	+20 + 4	+24
	5	8	50 + 8	58

	TENS	ONES		
	4	0	40 + 0	40
	+ 2	9	+20 + 9	+29
			+	

	TENS	ONES		
	1	5	10 + 5	15
	+ 3	3	+30 + 3	+33
			+	

TENS	ONES	TENS	ONES	TENS	ONES	TENS	ONES	TENS	ONES
6	2	2	6	8	1	4	3	5	6
+1	7	+5	2	+1	6	+3	0	+2	3

TENS	ONES	TENS	ONES	TENS	ONES	TENS	ONES	TENS	ONES
4	5	1	4	5	0	7	7	8	3
+3	1	+7	3	+1	8	+2	0	+1	5

24

ADDITION

ADD AND SOLVE THE RIDDLE.

66	79	46	47	26	69	57	87	55	98	39
B	E	G	H	I	M	N	O	S	T	W

When will water quit running downhill?

14 + 25	36 + 11	55 + 24	24 + 33		14 + 12	57 + 41
39						
W						

34 + 12	33 + 46	26 + 72	41 + 14		84 + 14	46 + 41

43 + 55	25 + 22	12 + 67

24 + 42	35 + 52	15 + 83	78 + 20	24 + 63	45 + 24

25

Write the equations and solve the problems.

Gavin and Justin went into The Sports Store. Inside, Gavin counted 56 footballs and Justin counted 23 basketballs. How many balls did the boys see all together?

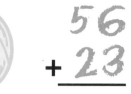

$$\begin{array}{r} 56 \\ + 23 \\ \hline \end{array}$$

Shelly and Marta went into The Candy Store. Shelly bought 46 pieces of candy. Marta bought 22 pieces. How many pieces of candy did the girls buy all together?

$$\begin{array}{r} + \underline{} \end{array}$$

In a large cage outside The Pet Store, Ryan saw 25 parrots. Inside the store, he saw 12 more. How many parrots did Ryan see in all?

$$\begin{array}{r} + \underline{} \end{array}$$

Julie had 37 pennies and Allen had 41 pennies. How many pennies did Julie and Allen have all together?

$$\begin{array}{r} + \underline{} \end{array}$$

Nancy and Kayla went to The Art Store. Nancy bought 63 paints for her paintbox. Kayla bought 34 paints. How many paints did both girls have in their paintboxes?

$$\begin{array}{r} + \underline{} \end{array}$$

Brian played with the train set in The Toy Store. The train he was playing with had 54 cars on it. He added 35 more cars to the train. How many cars were on the toy train?

$$\begin{array}{r} + \underline{} \end{array}$$

ADDITION

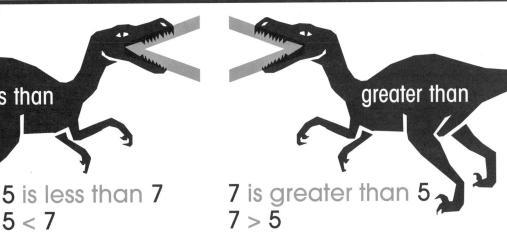

less than

greater than

5 is less than 7
5 < 7

7 is greater than 5
7 > 5

Remember: The dinosaur takes the greater number.
Put the dinosaur's "mouth" in the circle.

82 (>) 62   34 ( ) 4   40 ( ) 80

83 ( ) 33   10 ( ) 17   2 ( ) 8

55 ( ) 15   71 ( ) 74   35 ( ) 75

39 ( ) 25   60 ( ) 50   75 ( ) 57

43 ( ) 13   20 ( ) 55   82 ( ) 83

27

EXAMPLE:

3 tens   30

+ 4 tens   + 40

7 tens   70

20 + 40 60	40 + 50	50 + 10	20 + 70	10 + 40
20 + 30	10 + 60	60 + 30	80 + 10	70 + 10

Write the numbers. Then add.

50

+ 20

70

+

+

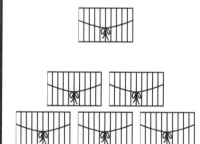

+

ADDITION

Add.   23
       + 42

First add the ones.

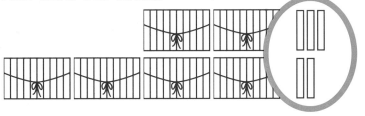

TENS	ONES
2	3
+ 4	2
	5

Then add the tens.

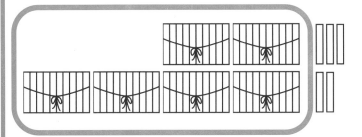

TENS	ONES
2	3
+ 4	2
6	5

TENS | ONES	TENS | ONES	TENS | ONES	TENS | ONES	TENS | ONES	TENS | ONES
35 +34	71 +18	15 +13	30 +56	17 +30	22 +75
24 +42	31 +10	12 +25	21 +37	46 +51	45 +10
41 +36	65 +33	40 + 9	11 +35	83 +16	77 +10

29

80 + 16	22 + 46	26 + 2

24 + 12	33 + 33	56 + 22

35 + 11	20 + 75	12 + 40	27 + 12	41 + 11	83 + 15
56 + 32	50 + 30	47 + 51	24 + 65	35 + 53	73 + 25
			24 + 42	23 + 74	42 + 35
			31 + 26	40 + 19	20 + 16

30

ADDITION

ADDING COLUMNS OF NUMERALS.

Add the first two addends.
Then add the last number to get the sum.

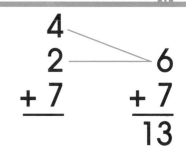

$$
\begin{array}{r} 4 \\ 2 \\ +7 \\ \hline \end{array}
\quad
\begin{array}{r} 6 \\ +7 \\ \hline 13 \end{array}
$$

$$
\begin{array}{r} 4 \\ 3 \\ +6 \\ \hline \end{array}
\quad
\begin{array}{r} 7 \\ 5 \\ +1 \\ \hline \end{array}
\quad
\begin{array}{r} 4 \\ 5 \\ +6 \\ \hline \end{array}
\quad
\begin{array}{r} 3 \\ 3 \\ +3 \\ \hline \end{array}
\quad
\begin{array}{r} 1 \\ 6 \\ +2 \\ \hline \end{array}
\quad
\begin{array}{r} 2 \\ 3 \\ +3 \\ \hline \end{array}
$$

Add the ones. Add the tens.

tens	ones
1	2
5	4
+ 2	1
	7

tens	ones
1	2
5	4
+ 2	1
8	7

$$
\begin{array}{r} 10 \\ 11 \\ +12 \\ \hline \end{array}
\quad
\begin{array}{r} 52 \\ 33 \\ +10 \\ \hline \end{array}
\quad
\begin{array}{r} 72 \\ 11 \\ +13 \\ \hline \end{array}
$$

$$
\begin{array}{r} 13 \\ 10 \\ +61 \\ \hline \end{array}
\quad
\begin{array}{r} 43 \\ 14 \\ +32 \\ \hline \end{array}
\quad
\begin{array}{r} 22 \\ 11 \\ +35 \\ \hline \end{array}
$$

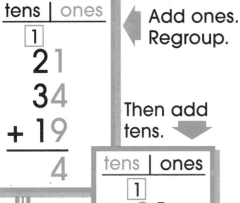

tens	ones
1	
2	1
3	4
+ 1	9
	4

Add ones.
Regroup.

Then add
tens.

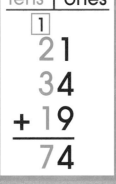

tens	ones
1	
2	1
3	4
+ 1	9
7	4

$$
\begin{array}{r} 23 \\ 25 \\ +24 \\ \hline \end{array}
\quad
\begin{array}{r} 14 \\ 42 \\ +36 \\ \hline \end{array}
\quad
\begin{array}{r} 66 \\ 12 \\ +17 \\ \hline \end{array}
\quad
\begin{array}{r} 43 \\ 25 \\ +26 \\ \hline \end{array}
\quad
\begin{array}{r} 29 \\ 15 \\ +20 \\ \hline \end{array}
\quad
\begin{array}{r} 54 \\ 22 \\ +17 \\ \hline \end{array}
$$

31

Add.
$$37$$
$$+ 25$$

First add the ones.

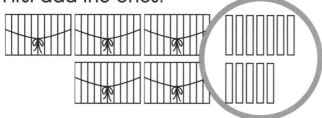

TENS	ONES
3	7
+ 2	5

Then regroup 10 ones as 1 ten.

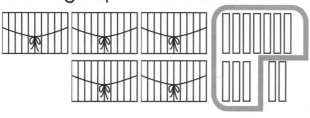

HINT: Think twelve ones!

TENS	ONES
[1]	
3	7
+ 2	5
	2

Then add tens.

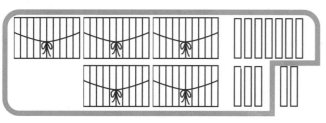

TENS	ONES
[1]	
3	7
+ 2	5
6	2

☐ 46 +46	☐ 68 +12	☐ 34 +28	☐ 55 +16	☐ 41 +19
☐ 36 +45	☐ 56 +25	☐ 29 +24	☐ 35 +29	☐ 14 +77

32

ADDITION

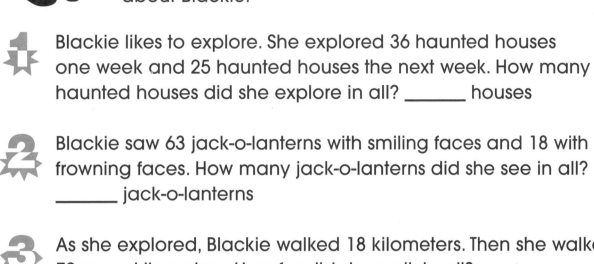

BLACKIE THE CAT

Blackie is a friendly cat. She doesn't believe she gives people bad luck by crossing their paths. Blackie lives with a friendly witch and a furry bat in a small house near the woods. Solve these problems to find out more about Blackie.

1 Blackie likes to explore. She explored 36 haunted houses one week and 25 haunted houses the next week. How many haunted houses did she explore in all? _____ houses

2 Blackie saw 63 jack-o-lanterns with smiling faces and 18 with frowning faces. How many jack-o-lanterns did she see in all? _____ jack-o-lanterns

3 As she explored, Blackie walked 18 kilometers. Then she walked 72 more kilometers. How far did she walk in all? _____ kilometers

4 When she got hungry, Blackie hunted 48 small mice and 27 birds. How many animals did she hunt in all? _____ animals

5 Blackie was full and tired. She napped in the sun for 55 minutes. Later she napped in the shade for 35 minutes. What is the total number of minutes that Blackie napped before going home? _____ minutes

6 The witch was so happy to see Blackie when she got home that she began to pet her. As she pet the cat, 81 hairs fell out of Blackie's fur. Later, 9 more hairs were lost. How many hairs fell out in all? _____ hairs

33

ADDITION

Add.
First add ones.
Then add tens.

☐
```
  34
+ 17
```

1
```
  34
+ 17
```
1

1
```
  34
+ 17
```
51

☐
```
  38
+ 26
```

☐
```
  65
+ 15
```

☐
```
  64
+ 28
```

☐
```
  48
+ 34
```

☐
```
  52
+ 28
```

☐
```
  37
+ 33
```

☐
```
  16
+ 16
```

☐
```
  23
+ 47
```

☐
```
  34
+ 18
```

☐
```
  15
+ 45
```

☐
```
  62
+ 19
```

☐
```
  33
+ 29
```

☐
```
  84
+  9
```

☐
```
  57
+ 26
```

☐
```
  16
+ 38
```

☐
```
  74
+ 19
```

☐
```
  21
+ 59
```

☐
```
  85
+  9
```

☐
```
  58
+ 26
```

☐
```
  36
+  7
```

☐
```
  34
+ 28
```

☐
```
  67
+ 14
```

☐
```
  29
+ 18
```

☐
```
  32
+ 49
```

34

FIND THE SUM.

$$\begin{array}{r} 76 \\ + 14 \\ \hline \end{array}$$

$$\begin{array}{r} 52 \\ + 29 \\ \hline \end{array}$$

$$\begin{array}{r} 14 \\ + 16 \\ \hline \end{array}$$

$$\begin{array}{r} 64 \\ + 16 \\ \hline \end{array}$$

$$\begin{array}{r} 34 \\ + 8 \\ \hline \end{array}$$

$$\begin{array}{r} 67 \\ + 3 \\ \hline \end{array}$$

$$\begin{array}{r} 67 \\ + 24 \\ \hline \end{array}$$

$$\begin{array}{r} 18 \\ + 14 \\ \hline \end{array}$$

$$\begin{array}{r} 58 \\ + 22 \\ \hline \end{array}$$

$$\begin{array}{r} 86 \\ + 7 \\ \hline \end{array}$$

$$\begin{array}{r} 27 \\ + 36 \\ \hline \end{array}$$

$$\begin{array}{r} 27 \\ + 44 \\ \hline \end{array}$$

$$\begin{array}{r} 19 \\ + 19 \\ \hline \end{array}$$

$$\begin{array}{r} 23 \\ + 49 \\ \hline \end{array}$$

$$\begin{array}{r} 62 \\ + 19 \\ \hline \end{array}$$

$$\begin{array}{r} 34 \\ + 58 \\ \hline \end{array}$$

$$\begin{array}{r} 36 \\ + 14 \\ \hline \end{array}$$

$$\begin{array}{r} 25 \\ + 48 \\ \hline \end{array}$$

$$\begin{array}{r} 66 \\ + 14 \\ \hline \end{array}$$

$$\begin{array}{r} 43 \\ + 38 \\ \hline \end{array}$$

35

ADDITION

18   **24**   **56**   **38**   **45**

$$\begin{array}{r} 38 \\ + 24 \\ \hline 62 \end{array}$$

   + _____

   + _____

 + _____

  + _____

   + _____

 + _____

  + _____

 + _____

  + _____

 + _____

 + _____

   + _____

 + _____

36

Shirley was white and had a shiny black beak and strong, beautiful wings. She knew how to swim in the pond and walk on the ground. Solve the problems to find out more about Shirley.

 1. Shirley's legs and body are 18 inches in height. Her neck is 14 inches long. What is her total height? _____ inches

 2. Yesterday, Shirley ate 36 kernels of corn. Then she ate 42 more. How many kernels did she eat in all? _____ kernels

 3. Last spring Shirley had 13 babies. A baby swan is called a cygnet. Another mother swan had 12 cygnets. What was the total number of cygnets born last spring? _____ cygnets

4. While in the pond, Shirley watched 45 big fish and 16 small fish swim under her feet. How many fish did she see in all? _____ fish

 5. Some children at the pond threw bread crumbs to Shirley and the other swans. At first, Shirley chased 31 swans away so she could eat the bread. Then she chased away 9 more swans. How many swans did she chase away in all? _____ swans

 6. It was turning cold and leaves were falling from the trees in the park where Shirley lived. 52 leaves fell from the oak tree near the pond and 27 leaves fell from the maple tree right next to it. How many leaves fell from those two trees? _____ leaves

ADDITION

FIND THE SUMS ON EACH STAR.

Star 1 (center: +26)
50

- 15
- 48
- 24
- 68
- 59
- 36
- 6
- 27

Star 2 (center: + 9)

- 16
- 45
- 32
- 12
- 68
- 5
- 21
- 54

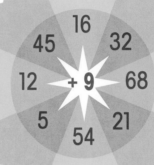

Star 3 (center: +44)

- 20
- 14
- 27
- 7
- 37
- 53
- 46
- 51

Star 4 (center: +13)

- 38
- 9
- 64
- 47
- 29
- 27
- 52
- 14

Star 5 (center: +32)

- 18
- 43
- 65
- 33
- 8
- 56
- 23
- 39

ADDITION

$§$OLVE THE PROBLEMS.

Write the answers in the correct circle.

Sums UNDER 50

Sums OVER 50

	23 + 24	18 + 18
12 + 15	46 + 35	30 + 15
14 + 16	17 + 25	26 + 12
19 + 57	63 + 36	33 + 9
59 + 36	23 + 46	16 + 66

39

20 →	7	
↓ 50	15	

30 →	15	
↓ 33	8	

26 →	14	
↓ 38	17	

18 →	21	
↓ 46	14	

37 →	16	
↓ 29	47	

49 →	54	
↓ 35	19	

40

ADDITION

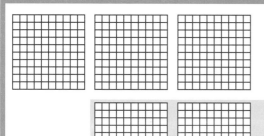

hundreds	tens	ones
3	0	0
+ 2	0	0
5	0	0

3 hundreds

+ 2 hundreds

5 hundreds

300 + 500	200 + 400	500 + 200	700 + 200	400 + 400
100 + 800	400 + 100	600 + 200	800 + 100	200 + 100

3 hundreds + 3 hundreds hundreds	5 hundreds + 4 hundreds hundreds	2 hundreds + 3 hundreds hundreds

Match.

600 + 200 •   • 700

400 + 300 •   • 300

100 + 200 •   • 800

41

First add the ones.

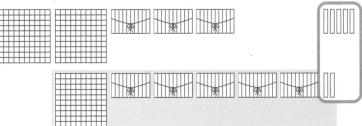

hundreds	tens	ones
2	3	5
+ 1	5	2
		7

Then add the tens.

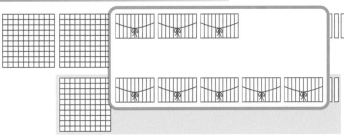

hundreds	tens	ones
2	3	5
+ 1	5	2
	8	7

Finally, add the hundreds.

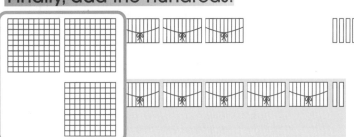

hundreds	tens	ones
2	3	5
+ 1	5	2
3	8	7

541 + 340	825 + 152	284 + 314	642 + 231	627 + 331	588 + 411	810 + 175
629 + 330	478 + 121	846 + 152	428 + 341	352 + 224	654 + 123	334 + 223

42

You may need to regroup 10 ones as 1 ten.

$$\boxed{1}$$
$$327$$
$$+516$$
$$\overline{843}$$

Regroup 13 as 1 ten and 3 ones.
Add tens. Add hundreds.

Add. On some problems you will need to regroup.

319 +532	144 +715	384 +213	643 +218	439 +236	549 +135
437 +201	126 +238	402 +285	542 +329	517 +426	213 +465
636 +215	328 +521	329 +529	233 +547	115 +358	659 +132
327 +641	337 +528	123 +546	239 +743	541 +340	449 +313

MARTIN THE MOOSE

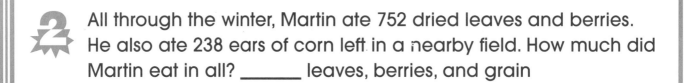

Martin is a big brown moose who lives near Yellowstone Park in the Rocky Mountains. He has thick brown fur, strong sturdy legs and a huge set of antlers. Solve the problems to find out more about Martin.

1 Martin weighs 317 pounds. His older brother Michael weighs 376 pounds. How much do they weigh all together? _____ pounds

2 All through the winter, Martin ate 752 dried leaves and berries. He also ate 238 ears of corn left in a nearby field. How much did Martin eat in all? _____ leaves, berries, and grain

3 Martin's antlers measure 60 inches across. His brother Michael's antlers measure 66 inches across. How many inches do their antlers measure all together? _____ inches

4 Martin measures 69 inches tall at the shoulder and his brother measures 78 inches tall. How many inches tall are Martin and Michael together? _____ inches

5 Martin fathered a baby moose called a calf. They named the calf Bobby. Martin is a great swimmer and he taught Bobby to swim. Martin swam 479 feet. Bobby only swam 127 feet. How many feet did they swim all together? _____ feet

6 Moose migrate in the winter months so Martin walked down from the mountains to the lower lands. He walked 127 miles one week and the next week he walked 158 miles. How many miles did he walk all together? _____ miles

44

ADDITION